Detective Zoro

Book 1

Mystery of the Haunted Lake

With Best Wishes!

Dear Reader,

Welcome to the world of Col. Zorawar, or 'ZORO' as he is known to his friends (& enemies!).

This is a world inhabited and ruled by DOGS! Dogs of all breeds, sizes, colours and temperaments. Take Col. Zoro for instance – he is an Indian Panther Hound. Then there is Simsim, who is a Mudhol hound and makes Col. Zoro's heart beat like a drum!

Dive right in and enjoy the tale of Col. Zoro's adventures – and remember, your heart needn't miss a beat because no matter how bad things seem, Col. Zoro will always set them right!

With love from the author!

Ranthambore Fort

This magnificent fort is built on top of a hill in Rajasthan. Surrounded by lush green forests on three sides, it was almost impossible to capture the fortress!

Join Col. Zoro as his first case takes him to this grand monument of India – a case that he can only solve once he reveals the true story behind the Ranthambore fort!

1 – One Private Eye

"I hope the **Coffeecue** is fine, Colonel?" asked the mongrel dressed in khakis.

"Yes Wodi, **perfect**!" replied Colonel Zorawar, smacking his lips in satisfaction.

Wodi, the mongrel, wagged his tail happily in response. He had been the colonel's *orderly*[1] all his life and knew all his likes and dislikes.

And like any good **army dog,** Col. Zorawar loved his barbecue chicken-flavoured coffee! It was a tricky preparation, which looked just like coffee but was actually a gravy made of barbecue chicken juices. But Wodi mastered it for the colonel, dishing it up for him through all his years of service in the army.

One Private Eye

Col. Zorawar had had a long and illustrious career. Known as 'Zoro' to all his friends, he had fought with and against all sorts of dogs, always **leading** his pack from the front. Be it with the Huskies in Siberia, the Alsatians in the Black Forest of Germany or the Cocker Spaniels in the British marshlands, Col. Zoro had seen action all over the world.

His favourite battle though was the one against the Chinese Chow Chows in the Northern Himalayas. He had bravely led the pack of Tibetan Mastiffs against the heavily armed Chinese dogs for a hard-fought **VICTORY.**

The fighting was so intense that Col. Zoro had lost his left eye in it.

He didn't mind that though. After all, like everyone told him, his **eye-patch** made him a very **dashing** *Indian Panther Hound*.[2] So much so that he had seven different eye-patches made – all of them in every colour of the rainbow, one for each day of the week!

"Well, Wodi, do we have any appointments for today?" asked Col. Zoro, looking up from his Coffeecue.

"None as yet, sir," replied Wodi, as he went back to **licking** the brass buttons on Col. Zoro's dress uniform for the army ball later that evening.

One Private Eye

Col. Zoro was a **detective** now. His stint in the army had trained him to **sniff** out the most complex mystery.

Once he found the trail, his quarry never escaped him. After serving his **COUNTRY** for so long, he decided that he would become a private detective to help other dogs in need.

"I am a Private Eye now. After all, I have only *one* eye and it is my *private* property," he liked to howl jovially.

So, there they were – Detective Col. Zorawar and his faithful Wodi, waiting for their first case...

2 – The General Acts Strange

Later that evening, Col. Zoro put on his dress uniform and headed to his old regiment mess. It was the regiment's annual day **celebration**, which is why he was in his dress uniform – the formal attire that all army officers wore on special occasions!

The party was hosted in the regimental ballroom, which was part of the mess. The place was beautifully decked up. All the old dogs were there, ***mingling*** with the current officers, as Coffeecue flowed freely.

"**Quite a party, General**," said Col. Zoro to General Sam, the old *Mudhol Hound*.[3]

Gen. Sam nodded absentmindedly. Then he suddenly asked, "Say Zoro, have you started your detective agency yet?"

"**yes, sir**," replied Col. Zoro, looking up from yet another cup of Coffeecue, as he licked his chops.

The General opened his mouth as if to say something but changed his **mind** at the last minute.

Col. Zoro peered at the General from over his mug.

There was something very strange about his ex-boss. Having served under Gen. Sam for many years, Col. Zoro knew something was very wrong.

"Is something the matter, General?" he asked with **concern.**

"Eh? What was that, my boy? No, no, nothing at all! Why would there be anything **MISSING?"** replied Gen. Sam, as he averted his eyes. "Hope you enjoy the ball, my boy. Have a bone while you're here. **WOOF**, woof for now!" Saying that, he hastily trotted away to greet the other officer dogs.

Col. Zoro looked back at his former commanding officer. 'Now why would he say that? I never asked him if anything was **missing**,' he wondered to himself as he began to gnaw on a **juicy bone**.

'Something is drastically **wrong**. I must keep an eye on Gen. Sam.'

General Sam made his way through the grand ballroom, exchanging pleasantries with the other guests. He did it so well that when he **quietly** slunk away from the venue, his tail tucked between his legs, nobody noticed… except Col. Zoro.

'This is very strange indeed,' thought the ex-army dog. 'The host leaves almost as soon as his party has begun – that too without informing anyone! Hmmm… maybe this is my first case as a **detective**.'

Dropping the bone in the tray of a passing steward, he quickly padded off in the General's wake.

"Let's see what Gen. Sam is up to," he said to himself.

"Maybe I can find out what's **troubling** him and help the old dog."

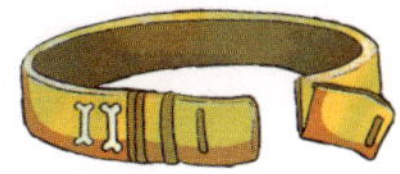

3 – The Golden Collar

Finding himself just outside the General's office, Col. Zoro heard hushed voices.

"I had WARNED you to be more careful," Gen. Sam's voice was loaded with worry. "I cannot keep this quiet for much longer."

Col. Zoro could see Gen. Sam but not whom he was speaking with.

"But I was very careful!" another voice wailed in response. "Oh, this is all my fault. What will we do now?"

Col. Zoro recognised the speaker immediately. It was none other than Simsim, the General's daughter.

Col. Zoro had known Simsim since she was a pup. In fact, they weren't too far apart in age and were very good friends. Col. Zoro had always had a **soft spot** for Simsim. Unable to contain himself any longer, he knocked on the General's **door**.

"**permission** to enter, sir," asked Col. Zoro.

"Come in," answered the General gruffly, inviting the detective to take a seat.

"What is it Zoro? We are a little **busy** here," he asked, sounding a little **CURT**.

"I am sorry, sir," Col. Zoro responded politely. "I didn't mean to intrude but when you **SUDDENLY** left the party, I got a little worried and followed you here. Even in the party, you looked a little distracted and not your usual self. Please tell me what the matter is, sir. Maybe I can help you. You know that you can **trust** me."

Gen. Sam looked at Simsim, who sat whimpering softly in the other chair. She looked up at her father with ***pleading eyes***.

After what seemed like a long time, Gen. Sam finally nodded. "Very well, Zoro," he said, "do you know about the Golden Collar?"

"Of course I do, General. If you remember, I'd won the Golden Collar when I passed out of the academy," replied Col. Zoro, his chest **swelling** with pride.

"Yes, you did now, didn't you?!" exclaimed Gen. Sam, a faraway look in his eyes.

Like any other officer, Col. Zoro too had started his career in the army at the *Officers' Training Academy*[4]. There, he had **trained** with the other dogs for two years.

Every year, amongst all the dogs that passed out of the academy, one was given the title of **'TOP DOG'**. And in his year, Col. Zoro had won the *prestigious title.*

The Golden Collar

The Top Dog was the dog who excelled in everything at the academy – physical training, fighting with all kinds of weapons, mental strength and EXCELLENCE in academics.

The Top Dog got a certificate and the coveted Golden Collar. While the Top Dog kept the original certificate, the same was not done with the collar. The academy took back the Golden Collar and gifted a replica to the winner to retain.

The original Golden Collar was a very precious relic. It was hundreds of years old and the most **powerful** and **inspirational** symbol of excellence in the army.

Its sentimental and symbolic value was far more than the price of the gold that it was made of! The Golden Collar was constantly kept under guard until it was time to felicitate the Top Dog for that year. Round the year, a unit of the most **ferocious** Rottweilers guarded it.

"Sim here has **LOST** the Golden Collar," announced Gen. Sam suddenly, triggering a fresh bout of whimpering from his daughter.

Col. Zoro looked up in shock.

This was **catastrophic!**

4 – Damsel in Distress

"But that is not possible!" exclaimed Col. Zoro. "The Golden Collar is never **removed** from the academy. Besides, there is no one that can even come close to it while the Rottweilers are on **guard**. How could Simsim lose it?"

"Tell him what happened, Sim," instructed Gen. Sam while his daughter used a lace handkerchief to wipe her tears. Col. Zoro's heart went out to poor Simsim.

"Oh Zoro, it is all my **fault**," she sobbed.

"**I should have listened to father**."

Damsel in Distress

"Calm down and tell me from the **beginning**," said Col. Zoro, trying to pacify the distraught girl. "What happened? Don't leave out any **detail**."

Gathering her wits, Simsim began, "Last week, I'd gone to the *Ranthambore wildlife sanctuary*[5] with a group of friends..."

"Yes, I know about Ranthambore," said Col. Zoro, feeling certain that his description of the place would be shorter than the one the General's daughter was about to offer. "It is a beautiful forest in Rajasthan, full of deer, tigers and even wild boar. They also have a **wonderful fort**, which is now in ruins." He then nodded, encouraging Simsim to continue.

"Since the passing out parade is just three weeks away, I told father that I would get the Golden Collar fixed while I was there," she started. "You see, the silver clasp that helps close the collar was broken. Udaipur, which is very close to Ranthambore, has **EXPERT SILVERSMITHS**. The plan was that I would take the collar to one of them and get it fixed."

Col. Zoro looked up in **surprise** at Gen. Sam. It was very unusual for something as precious as the Golden Collar to be given to a civilian, even if she was the General's daughter.

Gen. Sam looked ***very sorry***.

Damsel in Distress

"I know, Zoro. I shouldn't have **entrusted** her with this responsibility," he sighed, "but in my defence, it's not like she was on her own. Four of our best **Rottweiler commandos** accompanied Sim and they guarded the Golden Collar round the clock."

"**Strange** that the Golden Collar could go missing from under the noses of the *Rottweilers*[6]. They are the most **ferocious** guard dogs in the army. Anyway, what happened then?" asked Col. Zoro, turning back to Sim.

Simsim began **narrating** her tale in earnest.

5 – A Muddled-up Tale

Simsim, Jenny the Alsatian and Rani the Pomeranian had taken the train and reached Sawai Madhopur station in Rajasthan. From there, an **army jeep** had transported them to a lovely camp just outside the **RANTHAMBORE FORT**.

The fort is constructed in the middle of the forest. **Adjacent** to its wall, the local army unit had cleared away a small area to **set up camp** for Simsim and her friends.

She was, after all, the **General's daughter**! Each girl had her own tent to stay in.

"Oh, you should have seen the tents, Zoro; they were EXQUISITE!" exclaimed Sim, as she recollected the décor.

"It had deep, soft rugs on the floor. There was a huge four-poster bed in the centre with an exquisite writing table at one end. The tents were pitched on a wooden platform. That and the rug kept the interiors warm even during the freezing nights.

And the bathroom..." she swooned, "ooh, the bathrooms were the best – beautiful tiled floors with the loveliest fireclay sinks!"

"Very good, Sim. Where did you keep the Golden Collar while at camp?" asked Col. Zoro patiently, trying to bring the conversation back on track.

"**Oh right**, the Golden Collar," said Sim, suddenly coming back to the sad matter at hand.

"There was a tent pitched in the centre of the camp. That's where the Golden Collar was kept," she explained. "Three Rottweilers kept **guard** outside the tent at **ALL TIMES** while a fourth slept inside, right beside the Golden Collar. But you see, the Golden Collar didn't get stolen from the camp."

"**Huh**? Then where?" asked Col. Zoro.

A Muddled-up Tale

"I mean it got stolen from camp but that is because we went to see the haunted lake in the Ranthambore Fort," replied Sim.

"A haunted lake?" asked Col. Zoro, **surprised** at this turn of events.

"Yes, you see when we returned from the safari, we decided to spend the evening having a **barbecue** and picnic at the fort. That's when the **BULLY KUTTA**[7] told us about the haunted lake. Oh, but you should have come for the safari, Zoro; the tigress was **SO HUGE**! And she had such a lovely coat too," started off Sim, again getting **sidetracked**.

Col. Zoro realised that this was getting nowhere.

"How much **TIME** do we have, sir?" he asked Gen. Sam even though he knew the answer.

"**Four days** at the most," replied the concerned General.

The passing out **parade** of the next batch of officers was four days later. They would have to find the Golden Collar in time for the Top Dog  **ceremony**.

"If not, I have to report its loss," Gen. Sam said heavily. "I'm afraid we might even have to **arrest** Sim if it is not found. Or worse, I may have to resign in disgrace."

This last statement brought on fresh bouts of whimpering from Sim, as she remembered the **predicament** she was in.

"That will not happen, General... not while I'm there," said Col. Zoro with a **DETERMINED** look on his face.

"I will never ***allow*** anything bad to happen to dear Sim."

Then suddenly realising what he'd said, he added, a tad embarrassed, "And you, of course, sir!"

"That is all very well, Zoro. What is the **PLAN** though?" asked Gen. Sam.

6 – Zoro Has a Plan

"To start with, sir, I need your permission to **retrace** Simsim's movements from the time she left from here to the time she **DISCOVERED** that the Golden Collar was missing," replied Col. Zoro. "And I will need all members of the earlier party to **accompany** me."

"The Rottweiler **commandos** are still in Ranthambore," Gen. Sam responded. "As for the rest, they can leave whenever you say."

"Very well, sir. Permission to leave at **0500 hours**!" said Col. Zoro, saluting the General.

"**Granted**!" replied the general.

Zoro Has a Plan

"Sim, you and your friends be packed and ready to meet Zoro at the station sharp at 0500 hours tomorrow," he **instructed** his daughter **sternly**.

Simsim nodded back with a **GRIN**. She was excited to be going on another *adventure* with her old friend Col. Zoro! After all, she too had a bit of a soft corner for the **big dog**.

The General felt some of his old vigour returning, now that his **trusted friend** and lieutenant was personally investigating this case.

SAWAI MADHOPUR
STATION

Zoro Has a Plan

Soon, the group ARRIVED at the Sawai Madhopur station.

"Did you know that this station was constructed by Maharaja Sawai Mansingh of Jaipur?" asked Jenny the Alsatian, as she thumbed through a guidebook.

Col. Zoro nodded absentmindedly as they alighted from the train and started walking out towards the parking lot.

Along the way, they saw that the station was festooned with beautiful paintings on all its walls. They were all scenes from the Ranthambore forest. Some had tigers while others had sloth bears and deer. They were all colourful and showed the lush greenery of the forest.

Zoro Has a Plan

"Hello, sir, are you looking for a **guide**? I can show you the haunted lake of the Ranthambore Fort!" said a voice as Col. Zoro stood outside waiting for their car to **arrive**.

He turned to look and standing before him was a Bully Kutta. The girls were still a little distance behind.

"**HAUNTED LAKE**, eh?" asked Col. Zoro, sizing up the **stranger**.

The big Bully Kutta, or Indian Mastiff, was taller than Col. Zoro, with **RIPPLING MUSCLES** underneath his clothes.

"Yes, sir, haven't you heard of **Hammira's curse** and the haunted lake of Ranthambore?" asked the Bully Kutta, **bowing** before the Colonel.

Zoro Has a Plan

"I can't say that I have. Anyway, I don't need a guide," replied Col. Zoro evasively.

Then suddenly, he remembered that Simsim had told him that it was a Bully Kutta that had shown her the haunted lake. "Say, my friend," he called out to the dog, "we are staying at the camp near the fort. Why don't you come by TONIGHT and tell me all about the haunted lake and if I find your story interesting, maybe we can go see it too."

"Yes sir, I will be there," smiled the Bully Kutta before slinking away in the crowd.

7 – The Ambush

"Here we are **finally**!" exclaimed Simsim as Jenny and she came out with their bags. "I don't see why you have to pack so many clothes for just a couple of days," she said, turning to Rani the Pomeranian.

Rani emerged, **puffing** under the weight of her bag which was as big as she was!

"Oh, you don't understand Sim," she sniffed.

"We girls need our things."

Simsim and Jenny rolled their **EYES**.
"Here, let me help you with this," offered Col. Zoro, taking the big bag from Rani and putting it in the *jonga*[8] which had come to pick them up.

The Ambush

The *jonga* bounced along as they drove towards Ranthambore fort and forest.

They passed the busy town of Sawai Madhopur along the way.

The streets were lined with shops selling colourful handicrafts. There were traditional clothes with little mirrors sewn onto them, colourful hand-knotted rugs and covers and even paintings.

Then, of course, there were the shops selling delicious local foods, with the aroma wafting to the hungry travellers as their *jonga* chugged along.

In a few minutes, they left the town behind and turned onto a *narrow road*.

The Ambush

This road had the dense Ranthambore jungle on one side and open fields on the other. Suddenly, the *jonga* slowed to a **crawl**.

"Don't stop or get off the vehicle," commanded Col. Zoro. He was riding **shotgun** next to the soldier driving the jeep.

A fallen tree **blocked** the path ahead. The only way around was a dirt path to the side, which went into the jungle before turning again and connecting with the main road.

Either they would have to take that path or **move** the tree aside.

"What do we do, sir?" asked the driver nervously.

"Nothing. We keep going very slow and wait," replied Col. Zoro. "This tree hasn't FALLEN like this, blocking the road, by chance. It has been deliberately placed here this way."

"Who would do such a thing?" exclaimed the driver.
"I mean, who even knows that you are here in Ranthambore?"

The Ambush

"I don't know about all that," replied Col. Zoro, "but I am sure that if we take the jungle path, we will drive into an ambush. If we stop to move the tree, we will get **ATTACKED** on the road."

The soldier nodded as Col. Zoro and he drew their **PISTOLS** and waited to see what would happen next.

Suddenly, they heard the sound of an approaching **TRUCK** from behind.

Col. Zoro glanced in the rear-view mirror and saw a pariah dog leaning out of the passenger window, a **gun** pointed at their *jonga*.

"**Get down girls**!" warned Col. Zoro, before turning to the driver and saying, "That truck will try to **ram** into us from behind. As soon as he is close, turn hard to the right and go as **fast** as you can onto the path through the jungle."

"But, sir, you said that there could be an ambush there," **protested** the soldier.

"Just do as I say!" **commanded** Col. Zoro.

Being used to taking orders from his superiors, the soldier simply replied,

"**Yes, sir, Colonel**!"

WO-O-OF

They could see the truck gathering speed as it neared their *jonga*. When it was so close that it seemed as if it would surely dash into them, Col. Zoro shouted, "NOW!"

The soldier immediately revved the engine, turned the steering hard to the right and the POWERFUL *jonga* lurched towards the jungle path.
Its 4,000-cc engine was powerful enough to get the jeep from 10 kilometres per hour to 60 kmph in a jiffy.

On the other hand, the truck was going too fast to stop. Even though the driver slammed down hard on the brakes,
he couldn't stop in time to prevent what happened next.

The Ambush

While the *jonga* escaped with only a glancing blow to its side, the truck rammed into a tree trunk with a loud BANG! Col. Zoro looked back and saw the truck toppling over while the tree trunk rolled to one side of the road and into a **ditch**.

"Now quickly **REVERSE** the *jonga* back onto the road and go *full speed* to our camp," ordered Col. Zoro hurriedly.

As he spoke, the group spotted a few **pariah dogs** on the jungle path ahead. They were running towards them, brandishing **knives and guns**.

The Ambush

The pariahs chasing them hadn't even reached halfway before the *jonga* was back on the road and speeding towards the CAMP.

"Oh my, that was CLOSE," said Simsim from the backseat.

"Not at all, my dear. We always had things under CONTROL," replied Col. Zoro suavely.

"Why are we going so fast suddenly?" asked Rani sleepily.

She had managed to sleep through the entire adventure! The others heard her and burst out laughing.

9 – Ground Zero

In the next few minutes, the group turned right again, into the jungle. The road now had the forest on the left side and the **towering walls** of the Ranthambore Fort on the right.

“These walls look so forbidding,” shuddered Rani, glancing up at the sheer ramparts of the fort.

“Yes. Beautiful, isn’t it?” replied Jenny. “Imagine how life must’ve been for the people that lived here – this beautiful fort as their home and the entire jungle as their garden!”

Rani simply made a **horrified** face at Jenny. She was perfectly happy to stay at home and not venture out. “For the life of me, I can’t imagine why anyone would want to go to the jungle of all places,” she muttered.

Col. Zoro ignored them and keenly looked around him for traces of any danger ahead. He was a mountain dog, born and bred in the Himalayan jungles. This forest was anything but scary to him!

Soon, they reached a small clearing in the woods. The tents were as the girls had left the spot barely a few days ago.

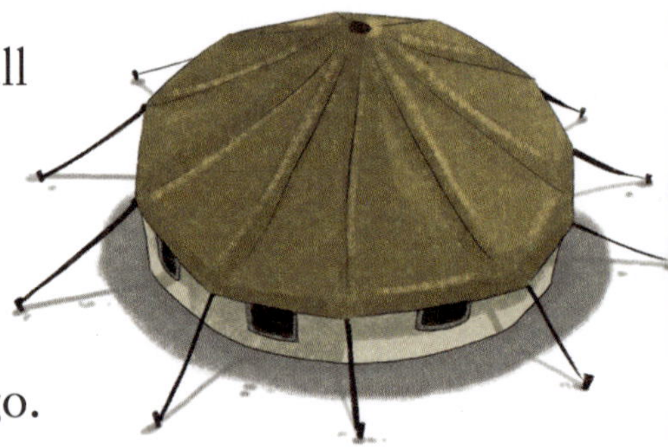

"Welcome, Col. Zoro!" said a Rottweiler wielding an automatic rifle.

He SALUTED smartly as Col. Zoro and the girls alighted from the *jonga*. "It is good to see you again, sir."

"At ease, Captain Tipu," replied Col. Zoro. He knew the Captain from his days in the army. He was known as a fierce fighter, extremely competent and loyal to the core.

After the Captain confirmed that his team was the one responsible for the precious cargo, the Colonel matter-of-factly stated, "I would really like to know how the Golden Collar could get STOLEN right from under your nose."

"Yes, sir, Colonel Sir!" shouted Captain Tipu. Though he held his head high, Col. Zoro could make out the PAIN in his eyes at having failed at his duty. His face was also lined from not having slept well since the incident.

Ground Zero

Col. Zoro nodded back at Captain Tipu with sympathy before turning around to look at the camp.

"So, this is where it all started," murmured Col. Zoro as he took in his surroundings. "The place where the Golden Collar went missing from."

"Yes. Horrible, isn't it, Zoro?" squealed Simsim. "The Golden Collar was taken away by the GHOSTS that haunt the lake in the fort."

Col. Zoro looked back at her, PUZZLED. She did tell him something about the Golden Collar going missing because of the HAUNTED LAKE, but nothing about ghosts having stolen it!

10 – Bully Kutta Returns

"What is this about ghosts, Captain?" asked Col. Zoro, as they all ***settled*** down in the evening after freshening up. They were seated on deck chairs around the unlit fire. The forest around them looked **beautiful** in the fading light and there was a slight nip in the air.

"Sir, do you know about *Maharana Hammira's* tale?" asked Captain Tipu.

"Of course. Haven't we all been taught about his valour and defensive strategies in the academy? But what has that got to do with **GHOSTS**?"

"Maharana Hammira...
WHAT A NASTY MAN!" shuddered Jenny.

"Yes. And what is this 'Maharana' business?" piped up Rani. She sat snuggled under a poncho with just her head peeking out, her ears twitching in the cold.

Before either Tipu or Col. Zoro could reply, they heard a voice from the jungle. Capt. Tipu immediately JUMPED up from his chair and pointed his rifle in the direction the sound had come from.

"*Relax, Captain*. I have been expecting this dog," said Col. Zoro as he put his **revolver** back inside his jacket's pocket.

Bully Kutta Returns

"*Khamagani* Zorawar *sa*," said the stranger dog in a reedy voice as he emerged from the **darkness** of the jungle.

"It's you again!" growled Captain Tipu menacingly. "I told you to **stay away** from our camp!"

"*Rana* Zorawar *hukum* called me here," said the stranger haughtily.

"You know him, Captain?" asked Col. Zoro with **mild interest**, as he sipped on his **Coffeecue** which Wodi had just prepared.

"Yes, Colonel. This is the scoundrel because of whom we lost the Golden Collar," growled back Tipu.

"Ah, I see! Well, come along and share our fire," Col. Zoro motioned to the strange dog.

Bully Kutta Returns

It was the same Bully Kutta that had **approached** Col. Zoro at the station earlier that day, offering his services as a **guide**.

"What was he saying – *Khama*-something?" whispered Rani as the Bully Kutta came and stood next to the fire.

"That is how you address someone with **RESPECT** in Rajasthan," explained Simsim as Wodi got the Bully Kutta a cup of **Coffeecue**.

"Here, '*Khamagani*' means 'greetings', '*Rana*' and '*hukum*' are similar to 'my lord,' while '*Maharana*' is a title for a '**great king**'."

"So, Mr. Guide, do you have something to say about the ATTACK on our *jonga* earlier this afternoon?" asked Col. Zoro, looking keenly at the Bully Kutta.

"I swear I have no idea what you are talking about *hukum*," replied the Bully Kutta. "I don't even know who would dare to attack you here!"

"Well, that doesn't matter. Tell me what can we do for you?" asked Col. Zoro.

"It isn't what you can do *hukum*; it is about what I can do for you," replied the BULLY KUTTA.

"And what is it that you can do for us?" Col. Zoro arched his eyebrows in mock **surprise**.

"Surely you are here to visit our magnificent fort, *hukum*. I, Batuk, am the best guide that you can find here to tell you all about the secrets of this fort – especially the **HAUNTED LAKE**!" replied the Bully Kutta.

"Oh, he is soooo right, Zoro," crooned Simsim. "In fact, when we last came here, Batuk took us on a tour of the fort. It was simply perfect. If only we hadn't had the Golden Collar stolen at the same time, this would have been our best **holiday** ever!"

"I see," said Col. Zoro, looking at the guide thoughtfully, "Very well, Batuk. When should we go on this tour with you?"

"It's a FULL MOON tonight, *hukum*. If we leave around 11 p.m., we will have a good three hours of moonlight in which I can show you the fort."

"Ah, that is perfect then. We have no other plans for tonight; 11 p.m. it is then," beamed Col. Zoro.

"You please rest, *hukum*. I shall return at 11," said Batuk, happy at having gotten his way.

Col. Zoro **NODDED** to Batuk before turning to Captain Tipu.

"Please have your men **escort** good Batuk out of camp so that he can see the **main entrance**. We don't want him returning by way of the jungle and getting shot by one of our guards after all, do we?" he winked.

Batuk laughed **nervously** as he was led away from the camp.

11 – A Twist in the Tale

"I don't **trust** that dog, sir," said Captain Tipu once Batuk had left.

"Forget about him. I want to know exactly how the Golden Collar got stolen right from under your nose," replied Col. Zoro **STERNLY** now.

"There is something sinister afoot here. First, the **same guide** who took you'll around the fort comes and meets me at the station. To add to that, our *jonga* is **attacked** on the way. Whoever is responsible for the theft knew we were coming here!"

Everyone **gasped** with shock as realisation dawned that Col. Zoro was right. It seemed like their every move was known to the enemy.

It was only by Col. Zoro's quick thinking that they had been safe so far.

"And remember – don't leave out any detail, no matter how small it may be," he warned Tipu.

Captain Tipu visibly paled and straightened in his chair. He had served on several dangerous missions with Col. Zoro and knew that he wasn't a dog to be trifled with. If he, Tipu, even tried to lie to Col. Zoro, he would find out in a jiffy.

Not that Captain Tipu had any intention of lying though; he was just nervous.

Never lie
to Col. Zoro!

"Sir, that day this Bully Kutta showed up, wanting to meet with the **guests**," began Captain Tipu.

"I met him and said that Miss Simsim and her friends were out on a **SAFARI** and would be back only later that afternoon. He told us that he would wait as he wanted to ask the ladies if they'd like a **GUIDED TOUR** of the fort.

Since he seemed like a local, we allowed him to wait till Miss. Simsim's group returned. Anyway, she had told us that she wanted to see the fort in detail."

Simsim nodded, **confirming** Captain Tipu's words. "Go on, Captain," instructed Col. Zoro.

"When the ladies returned, we brought Batuk to them," Captain Tipu recalled. "He recommended that they should see the fort at night as that way they would be able to truly enjoy the legend of the haunted lake. They agreed and he returned to call on them at **2300 hours** that night."

"Where was the Golden Collar all this while?" asked Col. Zoro.

"Safe under our watch. No one had even gone close to it," assured the Captain. "Once Batuk returned, I deputed two Rottweiler commandos to accompany them on the tour. It didn't seem safe to send the ladies alone with him at night."

"Of course. Did Batuk object to this **arrangement**?" asked Col. Zoro.

"Not at all, sir. If anything, he seemed very **happy** with it. 'It will make my job of guarding our esteemed guests easier,' he had said."

"**Interesting**. What happened next?"

"Sir, according to the report from the guards that accompanied them on the tour, Batuk gave them his version of the *history* of the fort."

"His ***version***?" Col. Zoro arched an eyebrow. "Yes, sir. A version where..." Captain Tipu began to answer before Col. Zoro ***interjected***.

"That is fine, Captain. I am sure I will hear his version **tonight**. I am more interested in what happened next."

A Twist in the Tale

"Well sir, they returned from the tour by 0200 hours," said Captain Tipu. "The ladies went to their tents whereas the commandos took their posts on **sentry duty**."

"What about the Golden Collar? When did you last **check** that it was safe?" asked Col. Zoro.

"At 0230 hours, sir. That was the last check. It was *safe and sound* till that time," replied Captain Tipu, averting his eyes from the Colonel.

"And what happened next?" asked Col. Zoro. His **sixth sense** told him something was amiss here.

"Er...ahem... sir, that is when the **ghosts** attacked the camp." Wodi gasped in **SHOCK**!

12 – Supernatural Visitors

"A ghost attacked your camp?" Col. Zoro asked with mild interest.

"Yes, Zoro, even we saw the ghosts," Simsim chimed in.

"It wasn't one. There were many of them and they all came swarming in from the forest. It was very SCARY!"

"I see. Please explain, Captain," said Col. Zoro sternly.

"Sir. At 0330 hours, we heard a sound from the jungle towards the east. Dismissing it as a deer or some other wild animal, we continued our vigil at the tent which housed the Golden Collar," began Captain Tipu.

"It was all my fault," sobbed Simsim, her muzzle covered with her paws as the tears flowed freely over her manicured claws.

"Your **fault**, eh?" asked Col. Zoro, intrigued. "How were the ghosts your fault, Sim? You didn't call them, did you?"

"Oh, if only we had ***listened*** to Batuk at the lake… none of this would have happened," Simsim continued **sobbing** as if she hadn't heard her friend at all.

"Can someone please tell me what's going on here?" said Col. Zoro, a tad **irritated** now.

"I'll tell you, Colonel," said Jenny. "While we were inside the fort, Batuk took us to a huge, beautiful lake. He told us how the fort was ruled by a

CRUEL KING, HAMMIRA.

One day, Alauddin Khilji – the benevolent Sultan of Delhi – attacked Hammira to rid the people of his cruelty."

"When Hammira realised that he could not defeat Khilji in a **fair battle**," she continued, "he threw all the **gold** and jewels that were there in the fort into this lake. He thought 'if I cannot **enjoy** this gold, then neither should Khilji'!

Eventually, Hammira lost and was **killed** in the battle. But Alauddin Khilji could never find the gold because Hammira had **THROWN** it into the lake.

Since that day, anyone **visiting** the lake has to put something precious into the waters, otherwise Hammira and his followers attack them at home and **steal** their most **precious possessions**," she finished, panting for breath.

Supernatural Visitors

"And that is why it's all my fault, Zoro," continued Simsim now. "I refused to throw any precious jewellery or money into the lake, calling Batuk's tale 'superstitious nonsense'. Because of this, that horrible Hammira and his ARMY OF GHOSTS attacked the camp and made away with the Golden Collar!"

Col. Zoro merely looked at Captain Tipu. The Rottweiler was a seasoned soldier and not one to get taken in by any attacker, ghost or otherwise.

"Sir, we saw these ghost-like CREATURES with glittering eyes coming through the trees," confirmed Captain Tipu. "We shouted a WARNING for them to retreat but they wouldn't listen. Eventually, I gave the order to fire, but it seemed like the bullets would simply go through them."

"And where was the fourth Rottweiler?" asked Col. Zoro **shrewdly**.
"Was he in the tent all this while?"

"Yes, sir – except for **five minutes** when he came out to help us battle the **intruders**," replied Captain Tipu.

"**Ha! Just as I thought**! And when he returned, he found the Golden Collar **missing**, is it not?" asked Col. Zoro.

"Yes, sir. But the strangest thing is that neither anyone came **near** the tent, nor was it torn or damaged in any manner!" replied Captain Tipu **indignantly**.

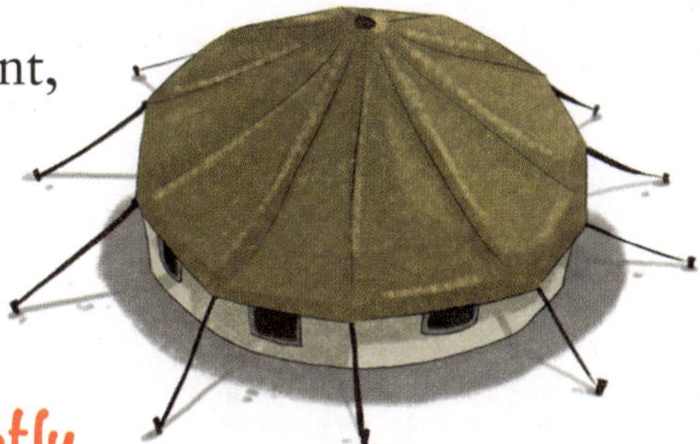

"Of course it wasn't, my boy. But not to worry. I am here now and all **shall be fine**!" assured Col. Zoro. "Now, let's get a bite of supper and some **rest** before we join this Batuk on a visit to the fort. Who knows, we may teach him something new about the history of *Maharana* Hammira and Ranthambore Fort today!"

13 – Ranthambore Fort and a Tale of Deceit

As promised, Batuk was back at 11 p.m. that night. This time, he came in from the main entrance. Col. Zoro and the ladies were ready and they left **immediately**. Captain Tipu had once again deputed one of the Rottweilers to accompany the party, to **guard** against any untoward **incident**. He himself stayed back at the camp with the other two **Rottweilers**.

"*Hukum*, this fort was made by the Chamana dynasty in the **10th century** CE," explained Batuk as they started ascending the **stairs** that led into the fort. The **massive** fort looked magical with the **moonlight** shining silvery on the stones.

A shudder went down Jenny's spine as she remarked, "**No wonder the place is haunted!**"

Ranthambore Fort and a Tale of Deceit

"The Chamanas ruled for 200 years over this region before the saviour Alauddin Khilji showed up. He was the Sultan of Delhi," continued Batuk. He then went on to narrate the tale which Rani had already told Col. Zoro, as to how Khilji attacked and killed Hammira, the last Chamana king.

"Interesting story, Batuk, but that is not how it happened," retorted Col. Zoro as they continued walking through the ruined fort.

"*Hukum*?" Batuk looked perplexed.

Ranthambore Fort and a Tale of Deceit

"The stairs that we **climbed**, Batuk," continued Col. Zoro, "Can you tell me why they **twist and turn** so much?"

Batuk merely shook his head in **response**.

"That is so that no elephants or horses can lead a **frontal attack**. Each time the elephants gather speed to knock down the doors, they have to slow down because of the **twisting stairs**. Likewise, horses and elephants are both very poor at climbing stairs, so they anyway have to go **slow**."

"That may be so *hukum*, but what has that got to do with the story of Alauddin Khilji defeating Hammira?" asked Batuk.

Ranthambore Fort and a Tale of Deceit

"Why, **everything**, my good dog, everything!" said Col. Zoro, before explaining, "You see, Hammira was the rightful king, the *Maharana* of Ranthambore. He was much **loved** by his people.

One day, two Mongol **generals** came to him, seeking **refuge**. They were serving in Khilji's army and had defected from there because Khilji asked them to do cruel and **un-soldier-like** things to the people they had **defeated**. Not wishing to unnecessarily torture their poor prisoners, the Mongols came to Hammira seeking **protection** from Khilji. Because you see, unlike what you have been telling everyone, Khilji was a cruel, unfair **cheat** who won the throne by **KILLING** his own uncle."

"After that, every battle he won was only by deceit and cheating, never by valour or bravery!" said Col. Zoro firmly.

Batuk merely put his head down and continued walking. The rest of the group looked at Col. Zoro, enthralled with this twist in the tale. They never imagined that Batuk had been lying to them all this while!

"Anyway, Khilji told Hammira to hand over the Mongol GENERALS to him, else he would attack Ranthambore. But the *Rajput*[9] code forbids anyone from turning away those that seek your protection and so Hammira refused," continued Col. Zoro. "Khilji sent several generals and armies to try and defeat Hammira, but couldn't succeed."

Ranthambore Fort and a Tale of Deceit

"Thanks to the **magnificent** defences of the fort, no one could even come close to the walls, let alone **breach** them!" declared Col. Zoro.

"That may be so, Zorawar, but Khilji did eventually **KILL** Hammira!" said Batuk rudely. In his anger at his lies being exposed, he referred to the Colonel **disrespectfully**, without using his title!

"Yes, you are right, but what is **IMPORTANT** is how such a thing came to pass," Col. Zoro continued, **pretending** that he hadn't noticed.

"Khilji bribed two of Hammira's generals to desert him with all the food and soldiers within the fort, leaving him almost defenceless. He promised them that he would make them the governors of Ranthambore after Hammira was defeated.

When Hammira discovered that his key generals had betrayed him, he realised that he had no choice. Rather than starve to death, he decided to take the fight to Khilji. So, Hammira, his brothers, sons and the few trusted warriors that were left, rode out of the fort to attack Khilji. The Mongol Generals joined them in this assault."

"What happened next, Zoro?" asked Simsim, mesmerised with this tale.

"They all got killed in the battle, Sim. They were too few against Khilji's massive army," replied Col. Zoro.

"And what of the TRAITORS? Did Khilji fulfil his promise?" asked Jenny.

"Of course not!" snorted Col. Zoro, "He was a man who was used to stabbing people in the back. As soon as he killed Hammira, he had the two traitors also put to death. Their names were Generals Ratipal and Ranmall. To date, their names are taken with DISGUST, for betraying a just and fair ruler like *Maharana* Hammira!"

12 – Mystery of the Haunted Lake

Everyone was so engrossed in this tale that they didn't realise that they had reached a beautiful lake in the middle of the fort. On one side were the ruins of a few structures while the other had the Ranthambore forest extending like a beautiful dark canopy under the moonlight.

"Isn't this amazing?" Col. Zoro remarked appreciatively.

"What – that this lake is haunted?" asked Simsim, shuddering with fear.

"No, my dear... that the Chamana kings managed to create this artificial lake at this height," replied Col. Zoro as he gestured to the forest far below.

Everyone looked around and realised he was right. The fort, and hence the lake itself, was at a far **greater height** than the forest!

"Before venturing out to attack the treacherous Khilji," continued Col. Zoro, "Hammira had all the **WEALTH** within the fort thrown into this lake. To this day, no one has found it."

"Yes, Colonel, you are right," Batuk found his opening. "To this day, the **ghosts** of Hammira and his **army** guard the treasure in this lake. All that come here need to put something precious in the lake, else their ghosts haunt them and take what is rightfully theirs," he said, baring his fangs in a **frightening smile**.

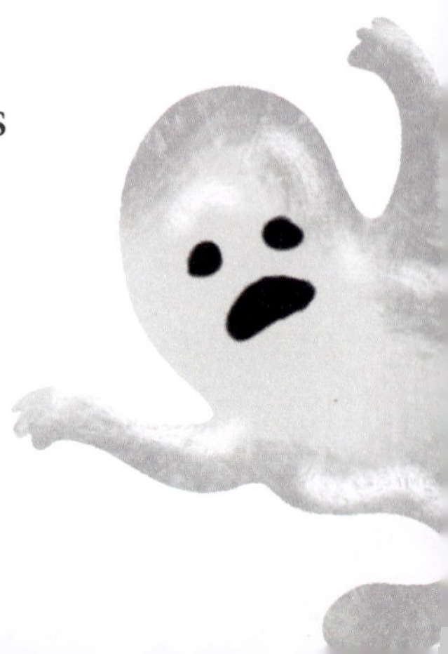

"I see, Batuk. So that is your story, eh?" asked Col. Zoro **matter-of-factly**.

"No story, Col. Zorawar, that is the truth. Please, for your own good, put something **precious** in the lake before you leave..." warned Batuk ominously.

"Oh, but I'm not carrying anything precious, you see," Col. Zoro **shrugged** his shoulders. "I suppose we should be on our way in that case. **I have seen all that I need to**."

Saying so, he turned to head back with the three girls.

"**Not so fast**, Zorawar," snarled Batuk.

Everyone turned around at the sound of his growl and were **SHOCKED** to see the Bully Kutta pointing a revolver at Col. Zoro.

"You can leave your most precious thing in the lake – **YOURSELF**!" he **growled**.

Col. Zoro noticed that the commando accompanying them lay **unconscious** on the floor. Evidently, someone had **WALLOPED** him hard on the head when he wasn't looking. 'Well, there is going to be no help from that quarter,' he thought to himself.

Just then, **two other dogs** slunk up beside Batuk.

Col. Zoro recognised them as the same dogs that had **attacked** them earlier that afternoon.

"Get into the water, Zorawar, and stay inside. **YOUR GAME IS OVER**," said Batuk. A low, satisfied growl came from the two pariahs.

"Oh no, you **villain**! What do we do now?" wailed Simsim.

"**Calm down**, dear. All is not lost," whispered Col. Zoro. He turned back to Batuk, "What is the **meaning** of this?"

"Why, I thought the meaning was **clear** enough ," replied Batuk, waving his gun menacingly.

"Get in the water," he barked. "But before that, my friends here will tie a ROCK to your paws so you sink and remain at the bottom of the lake.

If not, I will kill your lady friends, starting with Gen. Sam's daughter here."

Col. Zoro looked crestfallen. He wouldn't have bothered about any attack on himself. But the ladies – he could not let any harm come to them. There was nothing to be done. He had to listen to Batuk.

"Very well, Batuk," he sighed. "You leave me with no choice."

Hearing this, the Bully Kutta pointed his snout up towards the moon and howled with delight!

13 – The Tables Turn

"Now would be a good time!" shouted Col. Zoro.

Suddenly, three shapes shot out from the dark, moving like wraiths. Two of them knocked the pariahs unconscious, while the third – who was none other than Captain Tipu himself – pounced on Batuk.

The Bully Kutta was no pushover. He was bigger, stronger and as well-trained as Captain Tipu. Even though the Rottweiler had the element of SURPRISE, he could only manage to disarm Batuk.

As they grappled with each other, BITING and sniping with their sharp fangs and strong jaws, it seemed that the Bully Kutta had the upper hand.

Just then, Col. Zoro stepped up and picked up Batuk's revolver. Using the butt of the gun, he smartly hit the Bully Kutta on the head, rendering him unconscious.

"So much for that," said Col. Zoro, tucking Batuk's revolver into his waistband. "Well done, Captain. Tie up these miscreants and let's get them back to camp."

The commandos quickly followed Col. Zoro's instructions and the trussed-up dogs were carried back to camp. Wodi greeted them with freshly brewed Coffeecue and a cheerful fire. Col. Zoro and the ladies gratefully accepted the much-needed refreshment as they all once again settled down around the fire.

"So, we know that Batuk and his **gang** of pariahs are responsible for the theft of the Golden Collar," declared Simsim as she sipped on her Coffeecue, "but we are nowhere close to **retrieving** the collar itself!"

"All in good time, Sim, all in good time," reassured Col. Zoro.

He caught sight of Rani and gently inquired, "I hope you are well, my dear? ***You look a little pale.***"

"Huh? No, I'm perfectly fine. Must be all the **excitement**," replied Rani, looking away to hide the expression on her face.

The Tables Turn

"Yes, of course, that must be it. Here, come **closer** to the fire," said Col. Zoro, helping Rani pull her chair closer to the fire. "Time to rouse our guests I think, Captain," he said next, looking over at Tipu.

"Yes, sir!" Captain Tipu **saluted** smartly before issuing a command to his men. They splashed some ice-cold water on Batuk and the pariahs, who came to coughing and sputtering for breath.

"Welcome back to the land of the living, Batuk. I trust you are **well-rested** and none the worse for the knocks you've received?" smiled Col. Zoro at the Bully Kutta.

The Tables Turn

Batuk glared back at Col. Zoro. He was **fuming** at the turn of events. There he'd been, howling with delight at having bested the famous Col. Zoro, and in no time, the tables had turned on him. Instead of Zoro lying dead in the water,
it was he – the great Batuk – who was lying **battered** at the Indian Panther Hound's hind paws.

"Before anything else, Batuk, why don't you tell my friends here how you produced the ghosts in the woods?" asked Col. Zoro jovially.

The thug didn't reply; he simply glared balefully at the Colonel.

"What happened – cat got your tongue?" chuckled Col. Zoro. "Never mind, I'll tell you how it was done. A simple enough **parlour trick** actually."

The Tables Turn

He continued as everyone listened with rapt attention. "These pariahs working for Batuk carried flashlights with cut-outs of ghosts pasted on the lens."

"Huh? What?" asked a **bewildered** Jenny.

"Simple enough Jenny. The cut-outs were made with black paper. When they put on the flashlight, the projected beam would be in the shape of the cut-out, which in this case was ghosts," explained Col. Zoro.

"But, sir, we were **continuously** shooting at them. Surely, at least one of them would have gotten injured!" exclaimed Captain Tipu.

The Tables Turn

"Were you Captain?" asked Col. Zoro, smiling as he sipped on his Coffeecue. "You were shooting at the shadows that were projected on the trees, not at the pariahs carrying the flashlights!"

"Yes, yes, obviously!" piped up Simsim. "You see, the pariahs would've been hiding behind the trees, projecting the light on the trees opposite their positions. So, all the while that you thought you were shooting at the ghosts, you were simply shooting at the shadows!" she completed excitedly.

Captain Tipu nodded as he understood.

"All that is fine, but all that we have found out is that Batuk is a crooked guide," continued Simsim. "We are nowhere close to finding the Golden Collar!"

"Aren't we, my dear?" smiled Col. Zoro.

"**What do you mean**? Do you know where the Golden Collar is?" asked Simsim in return.

"**Of course, I do**!" replied Col. Zoro.

"**What**?" Simsim and Jenny exclaimed in unison.

"Where is it then, sir?" asked Captain Tipu.

"It is right here, Captain – **safe and sound** in this very camp!" Col. Zoro announced with flair.

A collective gasp went up around the fire.

14 – The Traitor is Exposed

"But sir, how is that possible?" sputtered Captain Tipu.

"Would you like to **enlighten** us, Batuk?" asked Col. Zoro, looking at the Bully Kutta.

"You were **LUCKY** to guess how we created the ghosts, Zorawar," growled Batuk, "but you will never find the Golden Collar. Soon, the whole world will know that you **LOST** the Golden Collar and your entire army will hang its head in shame.

Imagine, an army that cannot **PROTECT** even its precious little collar, how can its soldiers protect the whole nation? You will lose face and be the laughing stock of the whole world!"

"Ah, I see. And I suppose your masters, the **Chinese Chow Chows**, will then come and take over the army?" asked Col. Zoro.

"Yes, they will, as they **deserve** to," snarled Batuk.

"So that is what this is all about!" retorted Captain Tipu **angrily**, growling at Batuk.

"Yes, the minute I heard that the Golden Collar was stolen, I guessed it had something to do with the Chow Chows," Col. Zoro stated.
"The collar by itself isn't very precious; what it **represents** is more important."

"The Golden Collar is a symbol of the **power**, **integrity** and **greatness** of our army. By stealing the Golden Collar, the Chinese Chow Chows were looking to discredit the army chief and all the generals. We would lose credibility in the **eyes** of every dog and that would make their task of defeating our nation that much **easier**!" said Col. Zoro.

"But where is the Golden Collar, sir?" asked Captain Tipu. "We know that it was taken from the tent but like I already told you, the tent is not **damaged** in any way."

"Don't you find it odd, Captain Tipu, that Batuk here tells you all a **yarn** about **HAUNTED** lakes and evil kings and the same night ghosts show up at your camp?" asked Col. Zoro in return.

"Yes, sir. Now that you put it that way, I see the **connection**."

"And why do you think that the ghosts appeared, Captain? They certainly didn't steal the Golden Collar. You yourself said that you made sure that they came nowhere near the tent where the collar was kept," prompted Col. Zoro.

"It was a **diversion**, sir!" Captain Tipu caught on. "While we were busy fighting off what we thought were **GHOSTS**, someone went into the tent and stole the collar!"

"**Bravo**, Captain!" applauded Col. Zoro before asking, "But tell me this – you were all guarding the **entrance** to the tent. How then did someone enter it, take the Golden Collar and get away without you **NOTICING**?"

"That is what I don't understand," replied Captain Tipu, frustrated.

"You don't **understand** because you are looking for a complicated solution, my boy," explained Col. Zoro, before turning to address the one dog who had suddenly gotten up to leave.

"**PLEASE DON'T GO**, my dear," he gestured emphatically. "This is the interesting part, one where you are the **main character**!"

All the eyes around the fire turned to the dog who'd just been put in the **SPOTLIGHT**.

"Huh? Rani – the main character?" asked Simsim, **confused** at this turn of events. "What do you mean, Zoro?"

"*Quite simple*, my dear," continued the Colonel as Rani sat down again, looking scared.

"You see, Rani the Pomeranian is **small** enough to have simply raised the flap and entered from behind the tent while NO ONE WAS WATCHING! Once she grabbed the Golden Collar, she left the same way while all the Rottweilers were busy fighting Batuk's 'ghosts'. I'm sure a quick search of her bag will reveal the Golden Collar."

Saying so, Col. Zoro turned to his orderly with a **FLOURISH**. "If you would be kind enough to do the **honours**, Wodi?"

“With pleasure, Colonel,” said Wodi, before saluting and turning to Rani’s tent.

“I didn’t do it on purpose,” sputtered Rani, shivering with fright and licking her lips repeatedly. “H… h... h… he made me do it. He said he would kill me if I didn’t,” she howled in fear.

Before anyone could think of a response, Wodi returned from Rani’s tent with her **oversized** bag. They could all see the Golden Collar inside, *glimmering* as the light from the fire danced off it!

"It wasn't FEAR that made you steal this precious collar, my little Pomeranian," said Col. Zoro, now looking very angry, "but greed and jealousy!"

Looking at Simsim and Jenny's confused expressions, he continued, "When Batuk approached Rani with his plan, she saw it as the perfect way to get rich and discredit your father. With Gen. Sam and you getting blamed for the loss of the Golden Collar, she would permanently get you out of the way."

"Jealousy? I don't understand!" asked Simsim, looking shocked.

"Oh, don't you?" snarled Rani, suddenly looking like a very different dog!

"You're always the centre of attention, the one dog that all the officers fawn over while the rest of us have to make do with crumbs. And why? Simply because your father is the General? That makes you better than the rest of us? I couldn't stand to see your self-righteous face for one moment longer.

Oh, if only this Bully Kutta wasn't STUPID enough to get caught," Rani yipped, baring her fangs all the time.

"Enough!" growled Captain Tipu. "With your permission, Colonel, can I send the prisoners to headquarters so that they may be tried for their crimes?"

"Please go ahead, Captain," nodded Col. Zoro.

17 – For Honour and Glory

A truck full of troops arrived from the nearby army unit and put Batuk, Rani and the pariahs under heavy guard.

"Well done, Captain Tipu. Thanks to you and your team's timely action, the NEFARIOUS plan of these villains was thwarted," said Col. Zoro as they finally got ready to leave.

"Thanks to us, sir?" asked the Captain, bewildered. "I don't understand."

"Simple, my boy. Rani was supposed to steal the Golden Collar and hand it over to Batuk. But your team and you kept the pariahs at bay so well and the camp so strongly protected, that they never had a chance to do the exchange!" declared Col. Zoro.

"So when I suggested to Gen. Sam that I wanted to **retrace** Simsim's steps to find out how the collar was **LOST**, Rani and Batuk saw it as their opportunity," he explained further.

"You **knew** about their plan all along?" asked Captain Tipu in awe.

"**certainly not**, my boy. I only figured it out once Batuk **invited** me for his tour. I knew all along that he meant to kill me so that Rani could safely hand over the Collar to him."

"And that is why you had us hiding beforehand in the woods near the lake! You knew that was where he would try to kill you and thus **EXPOSE** himself!" finished Captain Tipu.

"That is right, my boy. After his attempt to kill me on the way from the station failed, that would be his only opportunity to do the job," Colonel Zoro beamed.

"Well, it is time for us to be on our WAY. We must return the Golden Collar to its rightful place before the passing out parade!"

Then, along with Simsim and Jenny, he boarded the train at Sawai Madhopur to return home. The Golden Collar was finally where it belonged, safe and sound AND with its silver clasp fixed as good as new!

"Well done, Zoro! Even though retired, you have saved the day for us yet again!" beamed General Sam. They were once again seated in the comfort of the General's office."

"Thank you, sir. It is my good fortune that my first case as detective was in the service of our army!" replied Col. Zoro modestly.

"It is a pity I cannot give you another medal in front of all the old dogs. But you know this mission has to be kept discreet," added Gen. Sam, looking a tad embarrassed.

Col. Zoro merely nodded back in response, as Simsim simply looked up at her saviour in admiration. All was well once again!

GLOSSARY

1. **Orderly** – A person who serves army officers. An orderly helps in preparing the uniforms, polishing the shoes, etc., and any other task that the army officer needs getting done!

2. **Indian Panther Hound** – It is a breed of dog found in the state of Uttarakhand in the Indian Himalayas. These dogs are related to the Tibetan Mastiff and are also known as Gaddi Kutta. This is because the Gaddi, a tribe of Himalayan shepherds, keep these hounds as sheep dog. Just 4 of these dogs are enough to manage 2,000 sheep! These dogs are extremely intelligent, incredibly brave, loyal, and great with kids.

3. **Mudhol Hound** – This is a dog breed found only in India. It gets its name from a place called Mudhol in the Indian state of Karnataka. Initially bred for hunting, these dogs are fiercely loyal, courageous, and very elegant.

GLOSSARY

4. Officers' Training Academy – Most countries have an Officers' Training Academy. Promising young women and men desiring to join the army as officers have to take a rigorous entrance exam, which tests them on their presence of mind, intelligence, academic knowledge and physical ability before they are admitted to the academy. Once in, they are trained to become excellent officers!

5. Ranthambore wildlife sanctuary – This is a tiger reserve near Sawai Madhopur town in Rajasthan state, which is in northwest India. Spread over 1,334 square kilometres, the reserve is home to tigers, marsh crocodiles, wild boar, leopards and deer. It is also home to a beautiful fort constructed in the 10th century AD, known as the Ranthambore Fort.

6. Rottweilers – Rottweilers are German dogs. They are stout, extremely strong and fearless. Belonging to the Mastiff family, the Rottweilers are considered the best and most ferocious guard dogs.

7. **Bully Kutta** – This is a huge, tremendously strong mastiff found in the Punjab region of Pakistan and some parts of Punjab in India. Called the 'Beast from the East', it is a naturally aggressive dog and was used for hunting.

8. **JONGA** – It stands for 'Jabalpur Ordnance and Gun Carriage Assembly'. This is the factory that used to manufacture this specific vehicle for the Indian Army under an exclusive license from Nissan.
The powerful jeep has a 4,000 cc petrol engine and can go virtually anywhere!

9. **Rajput** – This is a term that refers to warriors from north India, specifically Rajasthan. Literally meaning 'son of a king', Rajputs are renowned for their valour in battle and a code of always fighting fair.

Hope you enjoyed the first – ever case solved by the irresistible Col. Zoro!

If you thought this was a tough scent to pick up, then wait till you get your paws on his next case, *Legend of the Drowned City.*

Now we all know how much dogs hate water, so you can imagine what a tough time he would've had in sniffing his way through this mystery!

Forthcoming Titles:

Legend of the Drowned City

Abode of the Indomitables

The Comanche Connection

Mystery of the Secret Chamber

The 36th Puzzle

Quest for the Greatest Treasure

Secret Trail of the Warrior King